Final Destiny

Hassan Rasheed

Final Destiny

For further information, contact:
ostaahmed@yahoo.com

ISBN
978-1-387-97610-2

Published in the United States of America
By Lulu.com

Final Destiny

Introduction

"And now we stand in the entrance to humanities death along with the collapse of much of life on Earth." Yet, there is confusion among some of us as to the validity of this statement. It is the position of this publication that this downfall is inescapable. This stance is taken after the search for and discovery of the kernel of the problem which will be explained herein. It will be shown that the march towards this inevitability started not too long ago. And now the voyage is about to end in a most unpleasant and sad way.

Final Destiny

Chapter 1: The Ways of Earth!

"The beauty of nature is best known in waves of silence and stillness" Angie Weiland – Crosby

Earth's Cyclical Nature

Earth is a most self-contained system. Hardly anything comes into its system from outside, with the exception of sunlight and the occasional meteor; rarely anything leaves it with the exception of a little reflected sunlight and an occasional molecule of air floating into outer space.

What is the general logic of Earth's cycles? They help us to explain how an atom or a chemical compound move through their environment, starting in one part and moving to another, only to return back to where they started with no or very little in terms of accumulation. There is an endless number of cycles taking place on and within the Earth right now.

The cycle that can be perceived most readily is the water cycle which begins in the oceans, lakes, rivers and on land — as water vapor ascending through the air and forming clouds. These clouds will, in turn, drift along until the moment they give in and fall in the form of rain, snow, or ice onto land to form streams, creeks, and rivers eventually emptying into lakes and oceans or else are absorbed by the land.

The water cycle takes a year or more to reach completion and there are other cycles that may take less time or much longer. For example, Earth's crust consists of tectonic plates of rock, sand, and mud moving very slowly and colliding with one another. Science tells us that at their collision points one plate with all its mud and sand can slip under another, producing extreme pressure and heat. This results in volcanic activity that throws molten rock from deep beneath up to the surface. The rock cools and over millions of years breaks down to mud and sand once again.

Carbon dioxide also exists in the oceans, air, and underground. There is a balance of carbon dioxide between the atmosphere and the oceans, as the atmosphere and the oceans exchange carbon dioxide according to their temperatures and composition.

Scientists like to talk about these as carbon dioxide cycles. A major cycle in this regard is the photosynthesis/respiration cycle of living matter in the biosphere. For plants, photosynthesis takes carbon dioxide and water to produce Oxygen, sugars and starches upon which many if not most of the living depend on for energy. With respiration, the reverse can be observed: Oxygen and carbohydrates are combined, releasing carbon dioxide and water once again.

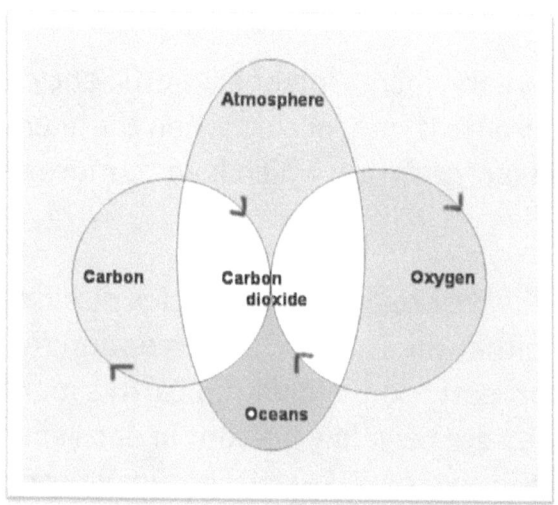

An example of the carbon dioxide cycles

The Living Cycles

In simplest of terms, life is a chemical reaction requiring energy. With a fixed amount of incoming energy from the Sun hitting the planet's surface, lowering the amount of energy required for this chemical reaction to take place allows for a proliferation of life. If the reader asks how they might lower the amount of energy required for a chemical reaction, the answer would revolve around the availability of catalysts and enzymes as a biological catalyst. In most cases where an enzyme is required, reactions occur faster because they require less energy to activate the reaction. You can think of a catalyst as a tool like a wrench which makes it so much easier to unscrew a nut.

In summary, increased energy efficiency using tools, such as an enzyme, occurs when there are mutations in our offspring which lead to more efficient enzymes.

As far as we are aware, the cycles discussed above are neither the rule nor the exception for the movement of matter on and within Earth's mantle. However, they are certainly present in each species in their circulatory systems. Specific groups of species— such as those found in a forest— circulate matter too. One can say that energy drives the circulatory system of forests, for example. Without it, forests would collapse because each of its links, whether one species or a piece of inanimate matter, provides a tiny link in the recycling of life-giving nutrients. This makes them available for regeneration of the forest.

When looking out a window or strolling through a park, the reader may observe trees, birds, grass, and bees, to name just a few of the life forms present in our surroundings. When focusing on a flower merely as an object, it may appear entirely independent of the other objects around it, but the truth is that it is intimately connected with all other objects in its immediate vicinity.

A clover, for example, needs soil to grow. As a mature plant, it later provides nourishment for the bison which roam the plains of the Midwest. The puma, meanwhile, later hunts out the bison for food. When it manages to bring one down, it drags what it can of the carcass to the shade of a tree to eat. Flies gather around the carcass to feed, lay eggs and try to avoid the birds that feed on them.

The above picture and explanation are obviously simplified, but they show how different species are so closely interlinked. A more accurate representation of the cyclical link between species is shown in the following graphic.

When following the biological cycle pictured above in a clockwise direction, starting with the green bush at the top left, a story can be told of how nutrients flow from one species to another. First, the bush extracts water and minerals from the soil and with the help of the Sun and atmosphere produces tender leaves which are consumed by the goat to build its muscle, maintain its growth, and hopefully also give birth.

The goat in turn is consumed by the human in the picture. The leftovers from the goat are used by the fly to lay its eggs. Some birds will feed on those flies, while the goat, human, bird, and fly leave behind fecal matter that is consumed by various microbes such as amoebas and bacteria, which return the nutrients back to the soil once again.

On any given territory, there are multitudes of life cycles, but perhaps with different species that participate in returning the atoms and chemical compounds (nutrients) back to the soil and into the atmosphere. In addition, these living cycles can often be interrelated. For instance, a living cycle that starts with a bush can have more than one grazer consuming it: they may be goats, sheep, or deer. These living cycles can take the following shape:

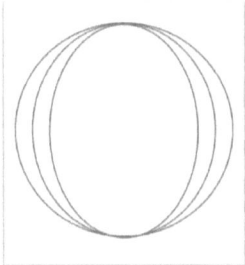

Looking at a square acre of land and trying to describe the number of living cycles within it, one may count many millions of combinations, their arrangement may look similar to the living cycles in the next diagram:

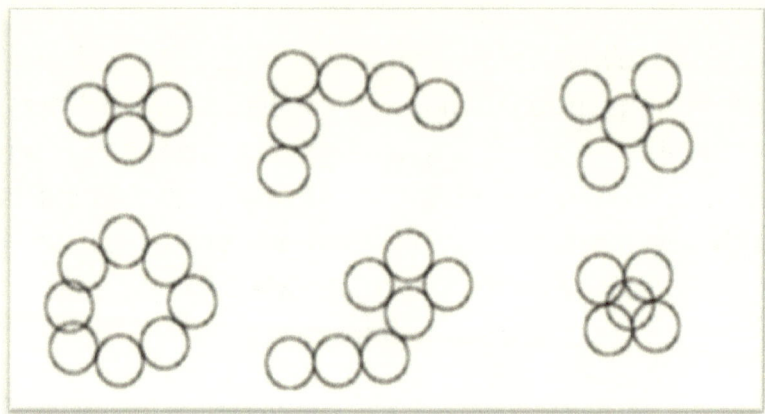

Living cycles are not limited to food, either. They can be observed in relation to nest building, territory, and so on. The biological cycle graphic below shows the cycle for a bird building a nest in a tree. When the eggs hatch, the nest may fall to the ground where insects break it down and pass on the resulting matter to amoebas and bacteria which will return the nutrients back to the soil to be used by the bush.

This is a key point in the discussion to introduce the concept of an ecosystem. A forest is an example of an ecosystem: Like any ecosystem, it contains mega cycles of circular chains involving both the air, soil, water, plants, and animals. A forest is not the only ecosystem. A desert such as the Sahara Desert is also considered an ecosystem.

Summary

The chain of life is circular, meaning that there is no start or end, no top or bottom link, nor is there any intelligent link or any particularly unintelligent one. The chain is nevertheless very complex. It contains many circular side chains. The chain is like a solid sphere where no one circular chain can be distinguished from another.

Final Destiny

Chapter 2: An Example of The Ways of Earth

The Forest People

(The Forest People, Turnbull, Colin M., 1962)

The Mbuti people live in the Ituri Forest, a tropical rainforest covering about 27,000 square miles of the north/northeast section of the Democratic Republic of the Congo, Africa. Mbuti are pygmy hunter-gatherers, and are one of the oldest indigenous people of the Congo region. They are composed of bands ranging from 15 to 60 people with a population hovering around 30,000 to 40,000.

Map #1: Location of the Ituri Forest in the Democratic Republic of the Congo

The Ituri Forest has a high amount of rainfall annually, ranging from 50 to 70 inches. The dry season is relatively short, ranging from one to two months in duration. The forest is a moist, humid region strewn with rivers and lakes. Disease is prevalent in these conditions and can spread quickly, killing not only humans, but major sources of food as well. Too much rainfall, as well as droughts, can greatly diminish the food supply.

In the rainy season they forage in part of the forest and in the dry season they expand their foraging territories. Their wild animal foodstuffs include crabs, shellfish, ants, larvae, snails, pigs, antelopes, monkeys, fish, and honey. The vegetable component of their diet includes wild yams, berries, fruits, roots, leaves, and cola nuts. While hunting, the Mbuti have been known to specifically target the giant forest hog which is often considered kweri, a bad animal which may cause illness to those who eat it, and is commonly used as a trade good between the Mbuti and agriculturalist Bantu groups. Folklore has it that the kweri giant forest hog is nocturnal and disruptor of the few imported agricultural activities the Mbuti practice.

Mbuti societies have no ruling group or lineage, no overlying political organization, and little social structure. The Mbuti are an egalitarian society in which the band is the highest form of social organization. Leadership may be displayed, for example, on hunting treks. Men and women basically have equal power. Issues are discussed and decisions are made by consensus at fire camps; men and women engage in the conversations equally. If there is a disagreement, misdemeanor, or offense, then the offender may be banished, beaten or scorned.

The forest is the center of Mbuti life. They consider the forest to be their great protector and provider and believe that it is a sacred place. They sometimes call the forest "mother" or "father". An important ritual is referred to as molimo. After the death of an important person in the tribe, molimo is noisily celebrated to wake up the forest in the belief that if bad things are happening to its children, it must be asleep and not taking care of them. As with many

Mbuti rituals, the time it takes to complete a molimo is not rigidly set; instead, it is determined by the mood of the group. Food is collected from each hut to feed the molimo, and in the evening the ritual is accompanied by the men dancing and singing around the fire. Women and children must remain in their huts with the doors closed. These practices were studied thoroughly by British anthropologist Colin Turnbull, known primarily for his work with the tribe.

The remoteness of the Mbuti people has protected them and their environment from the outside world for the most part in pre-1960. Their birth and death rates are high and roughly equal and as a result there are no overpopulation problems. Modern advances such as medicine and other technologies have had a rough time reaching them. Although there is no formal constitution or government, the context of their being is in their belief system: They are one with the forest. This factor has helped the Mbuti people escape the non-cyclical problems of other "modern" and "advanced" peoples such as deforestation, over use of resources, pollution, overpopulation and the accumulation of toxic wastes to name just a few difficulties.

Summary

If there is a living example of a viable goal post for humanity to live by on a healthy planet Earth, it is represented by the Mbuti people. They are not alone.

Other similar tribes exist today in remote parts of the Earth that exhibit similar characteristics. All deserve protection from the modern civilized societies.

Final Destiny

Chapter 3: The Kernel of the Planet's Problems

I have come to understand the human brain and its two most basic limitations: the first being a physical limitation in the quantity and quality of information it is able to store; the second is a result of the limitations of its sensory organs in perceiving the truth of nature. The brain is, nevertheless, a remarkable feat of engineering, with approximately 100 billion neurons each firing at a rate of once every 3 seconds. On each occasion that one specific neuron fires a signal, this information will be received by 7,000 other neurons. Multiply these three figures, and you have 200,000,000,000,000 bits of information transmitted every second inside your brain; a figure whose proportions elude the imagination of most. My point: The brain is limited in its ability to record all of this information.

With respect to the sensory organs, the eye can only perceive a very limited band of light, known as the "visible spectrum." The same sort of limitations exists in relation to time, sound, touch, smell, and hearing. Therefore, the brain is not able to detect the full scope of nature surrounding it.

For example, if the brain can store 50% of the information that is out there in the world and my sensory organs can process only 50% of what reaches them then the total product of information a person has is 25% of what is natural.

Although our brain's memory is reduced in information, the brain is able to manipulate information and even produce it. This is perhaps best demonstrated in how we solve problems. For example, by bringing together two different memories and producing a third that may solve some need. Unfortunately, since memories are limited in information and the process of linking two of them follows no natural law, the end result is an abstracted and artificial solution that will not necessarily prove to work in the natural world.

I now realize that the structures of the brain and those of the natural world are quite different things. Let us imagine that an understanding of the artificial world created by the human brain comes in the form of square blocks, with natural counterparts coming in the form of a ball. We might well imagine the blocks of our imagining to tumble and fall if built on the round natural world. When building a square structure on a round one, it immediately becomes unstable.

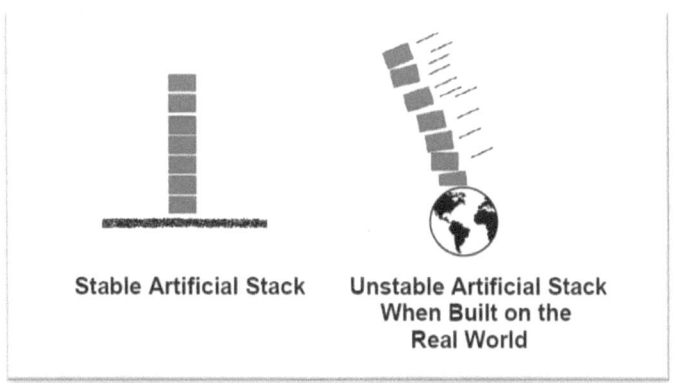

Stable Artificial Stack   **Unstable Artificial Stack When Built on the Real World**

According to such logic, anything humans can conceive or build is ultimately doomed to be short-lived. We should rather try our best to look to the structures built by nature which inevitably prove sounder and more durable than anything we could have built. In whatever we do, we should let nature set the path because nature is the final authority on what is right and suitable.

The Ego

A characteristic of most people on the planet is their misplaced ego. An ego marked by the cruel mementoes taken from their conquests such as the heads of the wild animals killed demonstrating their mistaken perception of power over nature. And so, killing of the natural took on a ritualized life of its own for the thrill of killing, acquiring trophies or building monuments to one's self. All demonstrating the arrogance and ignorance of the true ways of Earth.

Summary

Humans are not playing with a full deck when it comes to the knowledge of nature. To add insult to injury, their egos are inflated with self-centered and un-natural ideas. This makes civilized humans very dangerous for the planet Earth.

Chapter 4: The Evidence

The Artificial People

The artificial people are those who turned their backs on nature and created a synthetic world of their own. Their philosophy was that humans were to own nature and use it to do their own inventive bidding. That the natural world was somehow beneath them. Some went as far as calling and believing the natural world a domain of sin, filled with debauchery, depravity and immorality.

The artificial people believed in a straight-lined world as opposed to one that was dependent on continuous cycles. They supposed the individual and the world had a beginning and an end. It also included a middle section fraught with toil and denial of natural impulses and urges.

They used two common behaviors found in the repertoire of natural acts known as cooperation and hording to cement the artificial people's beliefs in their philosophical ideas. These beliefs lasted for the most resent five thousand years or so.

In nature, cooperation was used by the indigenous tribes to increase the ability of the family unit to acquire resources such as food, clothing and shelter. This in turn allowed the family the ability to secure the survival of its offspring and therefor the next generation.

Since the arrival of the artificial people, cooperation and its surrogates of ultraism and sacrifice were touted as a sign of supreme benevolence. This elevated cooperation to the position of serving not only the family and tribe but also the town, city and nation as populations grew.

Cooperation evolved to become a technique of encouraging the artificial people to work together more efficiently and in greater numbers. This made them able to force nature into the provision of resources needed to keep feeding, clothing and sheltering the resultant growing artificial masses.

Natural hording is limited by the physical limitation of the species. For example, if you can only carry 20 bananas in your arms at any one time, then that is your natural limit. With the advent of the artificial people, the merger of cooperation and hording in the last 150 years or so into the notion of capitalism, lifted these natural limitations with the help of technology.

As a result, capitalism erected monuments to itself and those who espoused it. These monuments involved the clearing or elimination of natural plant and animal species from large land sections. Some lands were then built up with massive sky scrapers. Other large free flowing natural lands were used to build hordes of sprawling smaller buildings that extended for a multitude of square miles. Still, other natural terrains were replaced with plantations consisting of vast rectangular or square fields containing a select few species of plants and animals seen as fulfilling the food and clothing requirements of the artificial masses.

The insatiable hording of profits demanded capitalism spend large sums of time and energy to fulfill this vision. In the end, the family unit had to be sacrificed. At first, fathers had to work long hours and as capitalism cut costs so did his earnings, requiring mothers to seek employment to make ends meet.

"I will make your lives much easier by providing all kinds of time saving gadgets freeing you up to spend more time with your families," capitalism responded. But it spoke with a serpent's tongue. These marvelous and sparkling thingamajigs also distracted the parents from their responsibilities. For example, some parents sought the glamor of moving up the ladder of success in these glittering corporations while others could not get enough of those spangled contraptions and in both cases, this pursuit required longer working hours leaving many families fatherless and motherless. And so, the nuclei of the artificial societies suffered.

Families were not the only entities that were hurt. Practically all other aspects of the world, near or far, were affected adversely. The ground, oceans and the air all deteriorated as well as countless species and ecosystems were degraded if not eliminated for the sake of the gadgets and the pursuit of capitalism.

The Ego Revisited

Predator killing, such as dolphins, wolves or pumas, is a case of one predator, namely people, killing off another predator to take their food or habitat. Predator killing also comes under the guise of wildlife management. In the north west of the U.S.A. they have killed off most cougars so that sports hunters can have plenty of deer or elk to kill.

Conservation and trophy hunting are both artificial acts that display shameful conceit and obliviousness of the facts. The truth is that we are all dependent on one another either directly or indirectly in the natural web of life on planet Earth. Conservation hunters take away from nature's ability to balance itself. We should treat our fellow species with dignity and respect.

Summary

As previously stated, humans are, in fact, incapable of grasping nature's full complexity. The truth is that humans are but one link of equal value to any other in the chain of life on Earth. The artificial people must rid themselves of the notion they are more intelligent or somehow superior to any other life form. The artificial people must learn humility and respect for the order and boundaries of nature.

Final Destiny

Chapter 5: Results from the Artificial

Current evidence indicates that the artificial people are returning climate conditions to their levels 60 million years ago when there were no humans and global temperatures were higher with a different atmospheric chemistry. There is no question as to whether the activities of the artificial people are driving this process by exhuming ancient fossil fuels. It is as if nature is pulling the artificial people by the nose to show them the extent to which their capitalistic systems, that created the current conditions, are unsustainable and need to be eliminated.

The Artificial Population Boom

The human population has exploded from a few tribes in Africa 300,000 years ago to a global population count of 7.5 billion today, predominantly due to our activities. Today, 96% of the biomass of mammals on Earth consists of the artificial humans, their pets, and farm animals. If the "ultimate achievement" of a species is to increase its numbers, we might ask at this point, "Are there any negative side effects to the apparent "success" of the artificial people?"

A density comparison between the Mbuti (forest people) and those of the artificial is that the Mbuti live at density of 1.3 people per square mile while the artificial live at a density of 122 people per square mile. The implication is that there is greater stress on the land and the people from the artificial people.

Forever Chemicals

Most people have some form of polyfluoroalkyl substances (PFAS) in their blood. While evidence indicates that exposure to certain PFAS sources is associated with human health problems, the available information focuses primarily on a small subset of 4,000+ compounds.

Populations are exposed to PFAS through ingestion, dermal contact and inhalation as a result of exposures to a variety of substances containing PFAS, such as food, water, dust, soil and consumer products.

Considerable epidemiological research has been conducted on PFAS in humans. That includes immunological, developmental or reproductive, hepatic, hormonal and carcinogenic effects. It is most certain that other species are equally if not more effected.

The Habitats

With the domestication of a limited taxonomy of plants and animals, more and more lands were required to support them, which resulted in a loss of habitat for the rest of the plant and animal species of no interest to the artificial people. For example, in the southern United States of America, beef cattle were raised on ranges where they competed with other indigenous species that naturally inhabited those areas. This competition was further weighted against them by the artificial population's tendency to kill by shooting or poisoning those undesirable species.

Soil Erosion and Degradation

Natural processes moved rocks from their resting places down hills and gullies through the power of wind and water currents, causing them to collide with each other. All these movements and collisions resulted in the formation of soils some 450 million years ago, when plants started to take root after evolving from much simpler organisms. Since then, soils have moved and developed through the very same processes.

In general, a "stable" soil is where this movement removes soil from an area at roughly the same rate as soil is deposited. A much more recent problem is the loss of soils at a much faster rate than their deposition. This is the result of artificial human action, such as overgrazing or inappropriate cultivation practices. These actions leave the land vulnerable and unprotected.

Soil may also be degraded by salination, nutrient loss, and compression due to other artificial people activities such as agriculture and construction. The result of soil erosion and degradation is a reduction in soil fertility and its support of living plants and animals, including the human species.

Deforestation

It has been estimated that approximately half of Earth's mature tropical forests, between 2.9 to 3 million square miles of the original 5.8 to 6.2 million square miles that until 1947 covered the planet, have now been destroyed. Some scientists have predicted that unless significant measures are taken, such as seeking out and protecting old growth forests around the world before 2030, there will only be 10% remaining, with another 10% in a degraded condition. 80% will have been lost and with them millions of irreplaceable animal, plant, and insect species.

Insecticides

Throughout most of human history, insects have been loathed. From locusts to termites and the weevil, they have gained a reputation as being detrimental to the interests of the artificial people. Recently, however, some insects were discovered to be rather essential. The bee is just one example. It wasn't until its beneficial effects on pollination were discovered that an appreciation of its services to the artificial people began to be recognized. Today, bees are in grave danger from the insecticides used by the artificial people to protect plants from other insects and fungi. Honeybees are critical for pollinating food crops. Scientists say the disruption of pollination could dramatically affect entire ecosystems. Yet, there are many more insects and insect larvae that are critical to human existence.

The Depletion of Global Fisheries

A report entitled *"Status and Solutions for the World's Unassessed Fisheries"* by the Sustainable Fisheries Group (SFG) has confirmed suspicions held by many researchers that nearly 80 percent of the world's fisheries are in steep decline. The reasons for this decline are overfishing caused by a multitude of factors, including ignorant political decision-making process, an unmanaged common, and technologies which improve detection and capture of marine wildlife.

Invasive Species

Invasive species are species that are transferred from one ecosystem to another, where they compete with native ones and take over their habitats. An increase in global trade has led to a further increase in the number of plant and animal species that are carried from one part of the globe to another legally, illegally, or accidentally. Some of these species become pests and wreak havoc on their destination ecosystems (such as rabbits and mice in Australia, for example, and blackberries in the Americas) often reducing biodiversity and threatening ecological balance.

Hazardous and Toxic Waste

Hazardous waste poses a substantial threat to public health and the environment. Worldwide, the United Nations Environmental Program (UNEP) estimated that more than 400 million tons of hazardous waste are produced each year by industrialized countries, with about 4 million tons shipped across international boundaries. The majority of these transfers occur between industrialized countries and developing nations where they are disposed of due to the rising cost of disposing of hazardous waste in the home country. Disposal consists of storage either above or below ground, leaving a legacy for future generations. In the case of biological toxins, they are often incinerated or simply released into the air.

Water Depletion

Groundwater is a valuable resource throughout the world. Where surface water, such as lakes and rivers, is scarce or inaccessible, groundwater supplies the needs of the artificial people. Sustained groundwater pumping as a result of overpopulation causes groundwater depletion and is a key issue affecting groundwater use. Many areas of the world are experiencing groundwater depletion. Pumping lowers the water table requiring drilling to greater depths, which in turn makes pumping more expensive.

In addition, the reduction of water in streams and lakes is partly due itself to groundwater depletion, reduction of the water table, and climate change. Landslides also occur when water is removed from the soil. As a result, the soil collapses and compacts.

The deterioration in the quality of fresh groundwater supplies is caused by saltwater intrusion from oceans, seas, deeper groundwater sources and the water below the oceans that are all saline. If we consider that one hamburger takes about 635 gallons of freshwater to produce, the problem becomes clear.

To look at water consumption on a different level, the average American household uses 107,000 gallons per day (2012). Rainfall and snowfall not only replenish ground water supplies but also provide for year-round water sources for many communities by replenishing snow to mountains in the winter that in turn provides water in the summer. However, our increasing usage of this rainwater due to water shortages proves a further problem.

Climate Change

Today we are witnessing the very early stages of global warming and its effects, but there are signs we are on the road to more significant changes in weather and ocean patterns. There are many agricultural communities dependent on the annual melting of glacier ice and its subsequent replenishment. In addition, the ocean and sea levels have been rising at the rate of an inch every ten years and are now accelerating.

In the United States of America, forest fires are burning larger sections of land and are reaching higher up the mountains where there was once snow to slow them down. These fires are also burning at hotter temperatures, and trees such as the Ponderosa Pines that depend on fires are also dying from the extreme heat these new fires generate.

Nitrous oxides play an important role in atmospheric chemistry. They are emitted into the atmosphere naturally, mainly as a result of microbial activities in soils and lightning discharges. Today, emissions predominantly occur as a result of **artificial people** activities (such as the combustion of fossil fuels, biomass burning, and the use of fertilizers).

Long-term trends of nitrous oxide concentration in the atmosphere are not documented adequately. Nevertheless, reconstructed emission inventories suggest that large increases have occurred throughout this century. Exposure to nitrous oxides has direct adverse effects on humans, animals, and plants. Nitrous oxides also contribute to the global environmental problems facing our planet (i.e., excessive global warming, depletion of the ozone layer, climate change and acid rain).

It is no secret that artificial people are stratified when it comes to starvation and death. Starting at a global level, the poorest nations bear the brunt of climate change and global warming through mass-famine and demise. And within each nation, whether the richest or poorest, the lowliest will experience the greatest hardship and the ultimate passing.

Plastics

Plastics is a product of fossil fuels. It is a convenient way to handle everyday products from cups that hold water to bottles to containers, wrappings and clothing. 8.3 billion tons of plastic has been produced. 6.6 billion tons ended up in landfills and the environment which includes land and waterways.

Few plastics get recycled—only about 9% in the US and 15% in Europe. The mechanical recycling that dominates today is hampered by contamination and a high variety in plastic waste streams. And those plastics that are recycled usually get turned into less-valuable products.

At least 14 million tons of plastic end up in the ocean every year, and plastic makes up 80% of all marine debris found from surface waters to deep-sea sediments. Marine species ingest or are entangled by plastic debris, which causes severe injuries and death.

Antibiotics and the Superbugs

Antibiotic resistance is a serious and growing problem in today's medical industry and has become one of the imminent public health concerns of this century. In the simplest cases, organisms acquire resistance through mutations that make them immune to common antibiotics, thereby requiring the use of less common secondary antibiotics.

Primary antibiotics are preferred due to their safety, availability, and cost. Secondary antibiotics are broader in spectrum, pose higher risks, and may be more expensive or difficult to source. In the case of some multiple drug resistant pathogens, resistance to secondary and even tertiary antibiotics have been demonstrated, illustrated by pathogens Staphylococcus aureus and Pseudomonas aeruginosa (which possess a high level of intrinsic resistance).

The main cause of wide-spread antibiotic resistance is the overuse of antibiotics by the artificial people in agriculture and in the health field. It is becoming harder and harder to fight infections caused by these drug-resistant pathogens and their increase in number and prevalence is outpacing man's ability to fight them.

Extinctions

We are living through a crisis in biodiversity. The fastest mass extinction in Earth's history is happening right now, largely due to the actions of the artificial people.

700,000 species have become extinct since 1976. This biodiversity crisis is snowballing and many experts predict a sixth mass extinction is on the horizon.

Summary

We are losing the race with time to save Earth's precious cargo. In 1993, Harvard Biologist E.O. Wilson estimated that Earth is losing about 30,000 species per year or 3 species per hour.

Today, climate change has grabbed the attention of most people through news headlines. But the large fossil fuel corporations are influencing the political systems against the transition to other sources of energy that do not produce greenhouse gasses. And as demonstrated here, climate change is only the tip of the iceberg of Earth's problems.

There is no doubt the artificial people are the direct cause of this habitat loss and the extinction of species through such events as over-population, the transformation of the landscape (through agriculture, construction, and uncontrolled forestry practices), overexploitation of species and their habitats, pollution, global warming, and the introduction of alien species into existing ecosystems. In other words, the self-proclaimed ideal species of the planet has run amuck.

Chapter 6: Recognizing Artificial People

It is not hard to do so. I know I am one of them. All you have to do is study my resume in appendix A and you will see why. In addition, I live with my wife in a wooden 900 square foot house on a clearing which once was an old growth forest. Gone are all the flora and fauna that once populated it. The wood that was used to build our house probably came from another old growth forest not too far way.

Our house has glass windows which means they were made from sand stolen off of some beach that once was inhabited by other species but are now all gone. All the plumbing fixtures and the nails that hold the house together had to be mined from giant gashes in the Earth's surface and again depriving more species of a home.

The furniture also includes metals and wood in their manufacture along with a host of other fixtures made of fossil oils such as foams and fabrics that go into the construction of sofas, beds and drapes. Fossil oils also go into the making of clothing, computers and audio-visual equipment.

The front yard has a huge willow tree. It is not native to our geographic area and therefor considered along with half the smaller plants and grasses in both front and back yards as invading species. The concrete pad that leads to the garage on the east side was also gouged out of a mountain side somewhere in the south of the U.S.A.

We own a car which further characterizes me as an artificial person. I do take solace in that it is a small 4-cylinder jobbie and therefore it caused less gashing of the Earth to build. In addition, its low miles per gallon rating shoots less nails in my sole than say owning a 6- or 8-cylinder gas guzzler. I bought a bicycle with the intent of further reducing my damage to the environment but I am up there in age and I eventually fell off of it and my wife forbade me from riding it again.

Like the majority of American families, we eat the typical foods found in the grocery store. And yes, we do go out occasionally and have a hamburger with fries and a cola. In other words, we consume edibles that produce the most damage to the environment than anywhere else on Earth.

We know of the cruel ways our foods are treated when still alive, yet as individuals we are powerless to do much that will significantly change anything about it. For example, slaughter houses and stockyards are not known for their gentle handling of livestock. And if my wife and I became vegan, the impact would be infinitesimal if any in stopping this type of immoral treatment. Genetic labs in mega corporations disregard the 4-billion-year wisdom of nature by creating fiends that help them turn a quick buck by altering life cycles of natural species.

A second glance at my resume will tell you that I had been a true in and out artificial person for most of my career because I worked in the manufacture of aero planes for an aerospace company. Not only did I support a manufacturing company that promoted further gouging of the Earth for raw materials but also produced the most polluting transportation vehicles on the planet.

Any heavy machine manufacturing company like that of the aerospace or car manufacturing industries own other regrettable processes. These processes use coolants, solvents, oils and coatings that are all harmful if not toxic to the environment. These poisonous ingredients either evaporate into the air, are discarded in landfills, incinerated or buried underground.

I am often asked, "Why don't you just go off into a jungle and live with the indigenous tribes?" I really did think of it but came to the conclusion that I had no right to do so because if I had I would be depriving one or two of the indigenous tribe people of a home. In other words, there just isn't enough room on this crowded planet for such a move.

I had been programmed into an artificial lifestyle from an early age. Examples of such indoctrination is the work ethic where I am encouraged to give it my all for 8 hours or more a day. To be punctual, smart and help the organization make as large a profit as I could.

We can't seem to take de-programming as a necessary step in returning to a more natural life style. We have become accustomed to our lifestyle with all its "trappings".

Other examples of artificial people are millionaires and billionaires. They have perfected the art of amassing wealth off the artificial systems. Nowhere in nature's handbook is there such a thing as affluence.

Other examples are the engineers, people who manage them and all others in between. We also can't forget the ordinary Joe who supports the above non-natural people in their pursuit of "happiness".

Summary

Most of the civilized are artificial.

Final Destiny

Chapter 7: Final Statements

So where do I start? Where should I begin? The issue is so vast! But Earth's pain compels me to write and to try to find an explanation. I am not to blame and yet I am responsible. I have departed from Earth's natural path and find myself with my head in the clouds, unable to see the reckless approaches in my advances treading on Earth's hallowed grounds in ignorance.

With Earth on the cusp of catastrophe, I have come to realize how we, the artificial, have somehow evolved to be this way. And I can't help but feel a deep sense of guilt and wonder as to whether the rest of the artificial peoples of this planet must feel just the same. After all, are we not, Sapiens, the "wise" ones who should learn to change our habits in order to accommodate Earth's ways.

The rising human population count has a huge impact on water, wildlife, forests, oceans, rivers, soils, atmosphere, migratory paths and just about every ecological factor you can think of. It is therefore imperative we allow nature to return the balance between birth and death rates for human populations.

Capitalism isn't freedom. Its premise is the accumulation of wealth through increased profit. In order to increase profit, it must cut down trees, gouge the Earth, acquire other specie's habitats or pollute the land, sky and water.

"Capitalism is serving us, the artificial consumers." There are two problems with this statement. First, capitalist companies are after efficient profitability where among other behaviors, cruelty is an acceptable tool. Second, artificial populations are encouraged to grow, consume and waste more to feed the profit mills. We need to abolish capitalism for the sake of restoring our humanity.

Nature is complex. Humans are, in fact, incapable of grasping its full intricacy. Humans belong to nature not the other way around. The artificial must rid themselves of the notion that "humans are at the top of the food chain" and "humans have superior intelligence." These are vocalizations of an ego desperate to force humanity's wrongfully perceived superiority upon others. We are but a link, no better or worse than any other, be it a clump of mud or another species in Earth's natural organization of life. The interference of our intellect and that of the artificial has fed our egos, leading us astray and blinding us to the consequences; the more we took, the less was left for others whom we are all dependent upon.

humans are incapable of forming any successful large society or economic system. We simply evolved to be tribal people meaning we are only able to form successful relationships within small groups. That is why we, the artificial, have so much history consumed in mega conflict and drama.

Indigenous tribes, like those of the forest people, are the salvation of the planet Earth. They live without damaging their environments. They have a nature based ecological footprint. It is criminal what is being done to them and their habitats. At the time they were researched in the 1950's by Colin Turnbull there were approximately 50 million of them living in the Congo River basin. Today there are less than 5 million.

I don't believe energy sources such as solar, geothermal, biomass, nuclear, wind, wave energy or other renewables are the solution. They only give us another way to gouge the ground for minerals, destroy more forests and obliterate the habitats of other species. We need a lifestyle that consumes much less energy so that we have no effect on Earth's ecologies and biota that we very much depend on.

I am sorry for what we have done to Earth by defiling its waterways, laying waste to its landscapes, and making its air the repository of toxins. I am sorry for my conceited attitude towards her. I am sorry the present and preceding generations of the artificial, who nurtured and took care of their own, did it at the expense of Earth's dignity. The artificial seemed a sure sign of success for a short time, but not indeed for long. And now, the artificial and the rest of Earth's species stand in the very doorway to the abyss.

There is one thing we must realize here and that is nature makes no promises. There are many examples of what are known as major extinctions in which a large group of species simply end up dead when their habitats no longer exist or are unable to support them.

In the past 540 million years it has been estimated that there have been five major extinctions when over 50% of animal species died. The following is a list of those extinctions:

1.   Cretaceous–Paleogene extinction event: 66 million years ago. About 75% of all species became extinct. In the seas it reduced the percentage of sessile animals to about 33%. All non-avian dinosaurs became extinct during that time.

2.   Triassic–Jurassic extinction event: 200 million years ago. About 70% to 75% of all species became extinct.

3.   Permian–Triassic extinction event: 251 million years ago. Earth's largest extinction killed 90% to 96% of all species

4.   Late Devonian extinction event: 375–360 million years ago. About 70% of all species became extinct. This extinction event lasted perhaps as long as 20 million years, and there is evidence for a series of extinction pulses within this period.

5.   Ordovician–Silurian extinction event: 450–440 million years ago. Two events occurred that killed off 60% to 70% of all species. Together they are ranked by many scientists as the second largest of the five major extinctions in Earth's history in terms of percentage of genera that went extinct.

The point I am trying to make is that major extinctions do happen and at the present time our only tool to predict that it is happening right now is science which tells us that it is most likely so. And now we must come to the uneasy subject of death.

Death indeed is a process all life on Earth experiences. The death we are facing here is that of millions of species and thousands of ecosystems. No matter how immense the event, it is a procedure of recycling which lays the foundation for new life. Nothing is really lost. For example, the Cretaceous–Paleogene extinction event 66 million years ago laid the groundwork for the evolution of humans and their cohabitant species. Atoms and molecules simply move around and eventually return to the land. By doing so, they get ready to brings forth new species and biomes with the help of the Sun.

As far as our intellect and consciousness are concerned, they will be deleted from the universe forever along with many vestiges of the voyage that got us to this tragic shore we find ourselves on today. It is necessarily so, simply because we rejected nature's ways and wrought against them with alien ideas. Ideas such as water shouldn't meander and that the secret to happiness on this planet is the accumulation of wealth and the spread of prosperity.

And so, from the soil will spring forth new species, ecologies, intellectuals and consciousness. Only nature knows of their kind. We can only hope that these new life forms are nature lovers for at least a while and realize that they are all an integral part of a fresh new world.

Appendix A

Final Destiny

RESUME
Hassan A Rasheed
ostaahmed@yahoo.com
www.rasheed.us

<u>Education:</u>
M.S. Applied Information Management from the University of Oregon, Portland Campus (1994) (Program in Web Systems and A.I.)
B.S. & M.S. (Biology) from the University of Oregon, Eugene, Oregon (1976) (Program in Evolution, Genetics and Ecology)

<u>Work Experience:</u>
2003-2004 Adjunct Professor of Software Engineering, Linfield College, McMinnville, Oregon
- **Software Engineering**
- **Programming Languages (Microsoft Visual Studio)**

1985-2003 Systems Analyst / Project Manager, The Boeing Company, Portland, Oregon
- **Project Management**
- **Systems Analysis**
- **Programming**

1982-1985 Programmer, Pearl Soft Inc., Wilsonville, Oregon

1980-1981 Peace Corps Volunteer, Kenya, Africa

<u>Published Works:</u>
- *A Case for the Genetic Adaptability for Sympatric Divergence*
- *Conditions for Strategic Alliance in the Large Commercial Aircraft Industry*
- *Nuraya*
- *Dear Desiree: Confessions of a Grandfather, autobiography*
- *An Introduction to Energy: Sources, Uses, Impact and Solutions*
- *Globus: The Economic Bases for Cooperation*
- *The God Equation*

<u>**Published Works Continued:**</u>

- *The Human Equation and the Prime Directive*
- *Earth Returned*
- *The Gods of Evolution*
- *Altruism: The Economic Bases*
- *Foreign Aid Madness*
- *After the Bloom: Earth's Salvation Manifesto*
- *In Search of Identity (historical novel)*
- *The Target for Our Ecological Footprint*
- *Deliverance*
- *Cataclysm by Capitalism*
- *Earth vs. Space: Your Choice*
- *Songs from Egypt*
- *Death of Anthony*
- *The Camel's Hoof*
- *Olon's Story*
- *Amazon*
- *Agents of Entropy*
- *Organic Philosophy and Climate Change*
- *How to Apocalypse*
- *The Modest Proposal*
- *The Grand Experiment*
- *Earth Apology*
- *Naiveté of Idealism*
- *Wrong Turn*
- *The Invisible Human*
- *The Price of Leaving Eden*

www.ingramcontent.com/pod-product-compliance
Lightning Source LLC
Chambersburg PA
CBHW020408290526
45785CB00005B/2467